HAND BOOK FOR LEARNING ALGEBRA

(A TEACHING LEARNING MODULE ON ALGEBRAIC REASONING FOR MIDDLE STAGE STUDENTS AND TEACHERS)

Prateek Chaurasia & Somu Singh

Made with ♡ on the Notion Press Platform
www.notionpress.com

Contents

Preface 7

Acknowledgment 9

About the Authors 11

Instruction for the Teachers 13

Instruction for the Students 15

Unit 1: Fundamental Concepts of Algebra **17**
    Introduction 17
    Instructional Scheme 17
        Polynomials 21
        Types of polynomials 22
        Degree of polynomials in two or more variables 23

Unit 2: Algebraic Expressions **25**
    Introduction 25
    Instructional Scheme 25

Unit 3: Algebraic Identity **31**
    Introduction 31
    Introduction to Algebraic Identities 31

Unit 4: Exponents **39**
    Introduction 39
    Instructional Scheme 39
        Laws of Exponents 42

Unit 5: Factorization I **45**

    Introduction 45

    Instructional Scheme...46

        Factors of Algebraic Expressions 48

Unit 6: Factorization II..**51**

    Introduction 51

    Instructional Scheme 51

Unit 7: Linear Equations...**59**

    Introduction 59

    Instructional Scheme...60

Unit 8: Applications to Linear Equations **63**

    Introduction 63

    Instructional scheme 63

Unit 9: Introduction to Graphs **71**

    Introduction 71

    Instructional Scheme...72

        Abscissa and Ordinate...72

        Quadrants 74

        Sign Convention on Four Quadrants 74

        Types of Representation on graph...........................76

        A Bar Graph 76

        Single Bar Graph 76

        Doubled Bar Graph..77

        Histogram 77

        Pie - Chart 78

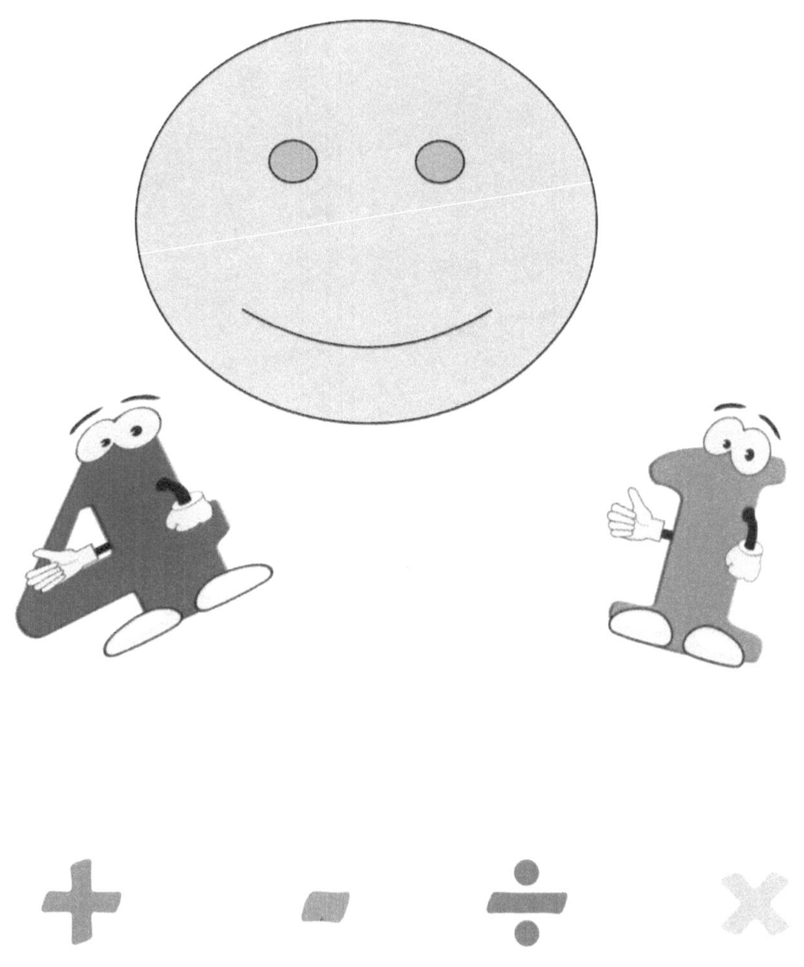

Learning algebra is fun

Preface

This teaching learning module on algebra is aimed to provide students and mathematics teachers an understanding of effective ways of learning algebra and enhancement of reasoning. Prime focus of module is to provide the various experiences of algebra learning to the learners. Since effective mathematics learning is highly depended on the activity that learners perform during learning the various concepts, the module emphasizes on equipping the learners with the fundamental knowledge of algebra as well as it contains the assessment at various stages and at the end of every unit of the module. Module is sub divided into the 9 different units based on the algebra and its related concepts. The present module is developed to improve the algebraic achievement and enhance algebraic reasoning among the class VIII grade students. This module contains the activity, exercises, reading materials along with the optimum opportunity to think algebraically. The prime goal of the present module is to develop the interest among the learners to learn the algebra. It covers the almost all the syllabus prescribed by NCERT and U.P. Board class VIII mathematics syllabus. With the help of module students will learn to identify, understand and appreciate the features, structures, and basic algebraic operations. The module also focuses on the development of the skills required to handle the various brain teasing exercises. This module is also helpful for mathematics teachers. This will equip them with easy graspable content and pedagogical approaches on complex and difficult areas of mathematics i.e. algebra. This module will also help them in making algebra learning joyful.

The units given in the module have especial features to link to the gap between the algebra and arithmetic and provide them a platform to advance their knowledge of algebra by using their prior experiences of arithmetic. Overall, joy full algebra learning is intended with the help of this module.

Acknowledgment

The module on algebraic reasoning is developed by very enthusiastic mentoring and the able guidance of Dr. Somu Singh (Faculty of Education, BHU). Along with him, I am also very thankful to Prof. Asha .K.V.D. Kamth retired professor of education R.I.E. Mysuru, NCERT. I extend my thanks to Prof. H.C.S. Rathore & Prof. R.P. Shukla, Ex. Head and Dean, Faculty of Education, Banaras Hindu University for their valuable guidance and suggestions regarding the development of this module.

My heartfelt and special thanks to Banaras Hindu University, Varanasi for providing me with the all necessary facilities and resources for the smooth and successful completion of research work and module development. I express my sincere gratitude towards the National Council of Educational Research and Training(NCERT), New Delhi, for providing me with financial support (NCERT Doctoral Fellowship) & opportunity of collaboration for the successful completion of this research work and module development.

Instruction for the Teachers

Teachers have the responsibility to provide a copy of the module to each student. After providing the module to the students, tell them to read the instructions given them in the module. It is the full responsibility of the teacher to ensure the peace full learning environment and discipline inside the classroom, while learning with the help of the module. Before starting the distribution of module to the students brief them with the basic purpose of the module and introduce the algebra among them. Instruct the students to think about the given activity and tasks within the module separately and do not allow them to take help of any other learning materials. Do not make any students in a hurry and let them take their own time while solving the questions and reading the module. During the module reading teachers have to play their own part as a facilitator, please do not provide any answers to the students and just help them in constructing their own knowledge by themselves. Teachers should provide all the resources available to ensure the participation of all the students in organizing the suggested activities in the module and also motivate the students to answer the assessment question given in the last of each and every unit of the module.

Instruction for the Students

There are nine units in the present module. All these units have been prepared to develop an understanding of the basic concepts of algebra and its related contents. You have to study each unit according to the instructions given or recommended activities in it. While reading units of the modules do not hesitate to ask something with the teachers. At the same time, try to solve the questions and exercises by using the hints and clues in the unit. Be sure to answer the questions given at the end of each unit and check the answers from your class teacher. Maintain peace and discipline during the study of the module. If no word or sentence is understood in the subject matter of the module, then definitely ask it with class teacher and start reading again. After completing the reading of units and module, inform the teacher.

Fundamental Concepts of Algebra

Introduction

As you all are familiar with algebra and its basic concepts like constant, variable, and its uses, but there are some other important fundamental concepts and operations like algebraic expression, and their degree. So, here in the present unit you will study about these all important fundamental concepts and operations of algebra.

Learning Objectives

In this unit, you will get an overview of fundamental concepts of algebra like variable, constants, and polynomials. In this unit attempt is made to discuss and elaborate all fundamental concepts and operations of algebra.

After working through this unit, you will be able to:-

- Recognize the variables and constants.
- Define polynomials.
- Calculate the degree of a polynomials
- Calculate the degree of polynomials in two or more variables.

Instructional Scheme

The main feature of this new branch (algebra) of mathematics, which you are going to study, is the use of *letter (x, y, z, etc)*. With the help of letters, we can write rules and formulas in a general way. In algebra main use of letters is to represent the unknown quantities. With the help of letters we get powerful ways to represent the unknown quantities and their relationship with other quantities (known or unknown) in the form of algebraic expression and other equations.

Some letters that we generally use in the study of algebra are:-

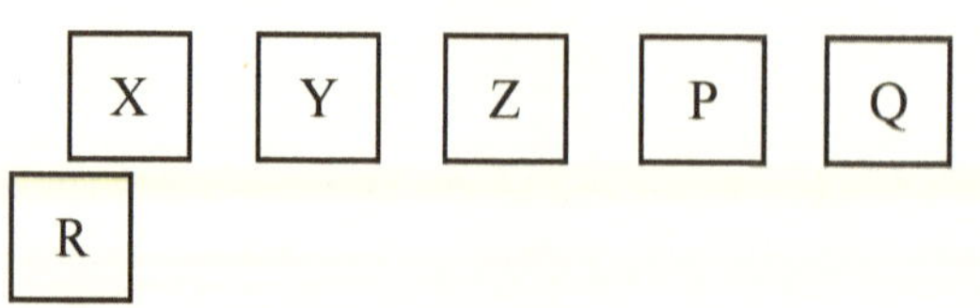

⚙ Activity 1.1

The height of Ajay is x cm and height of Praveen is 152 cm, but height of both Praveen and Ajay is equal to each other. Can you tell what the height of Ajay is?

Note: - Here x is used to represent the unknown value i.e. height of the Ajay.

Constant

A symbol in algebra having a fixed value is known as constant.

Example: 2, 7, 6 etc.

Variable

A symbol in algebra which can acquire different values is called a variable.

Examples: x, y, z, etc. all are variables

So, you have discussed about constant and variables but to know their properties uses and identification let's read ahead.

Let's see here, 2 is a real number has a face value = 2 and numeric value = 2 so, the value of 2 is fixed means it can only acquire a fixed value i.e. 2. Therefore it is termed as a constant

- Remember, a constant is a quantity which has a fixed value.
- Remember, a variable is a number which does not have fixed values.

⚙ Activity 1.2

Read and write in the blanks.

- How many hands you have?
- What does the number in the box that you have written represents, whether constant or variable?……………………………………..
- Number of questions in each exercise of your chapter is?…………………………….
- What does the above number represents constant or variable?…………………….
- Height of Ajay is 'x'. Here x represents a ……………….?

Fill in the blanks

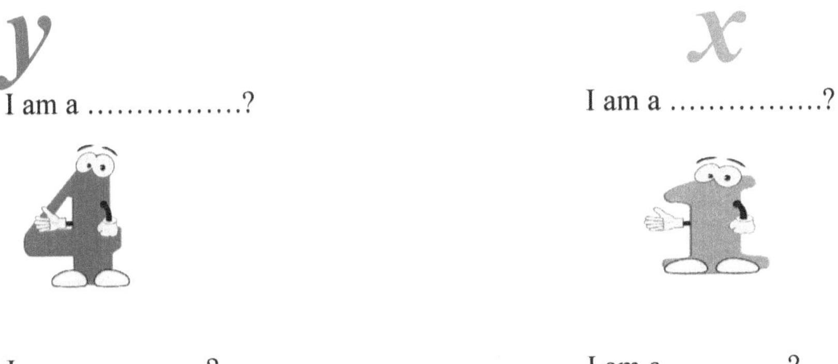

y

I am a ……………?

I am a …………?

x

I am a ……………?

I am a …………?

⚙ Activity 1.3

Ravi is interested in arranging the tables in his house for his guest, but he is not sure that how many guest will come to his house. So he tried to arrange the tables as shown below.

Arrangement 1

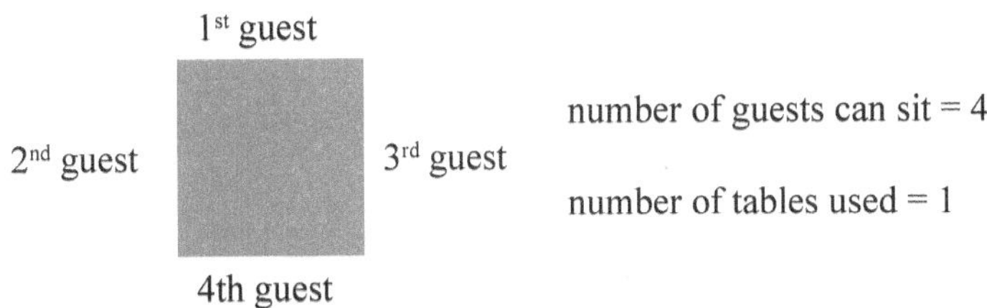

Arrangement 2

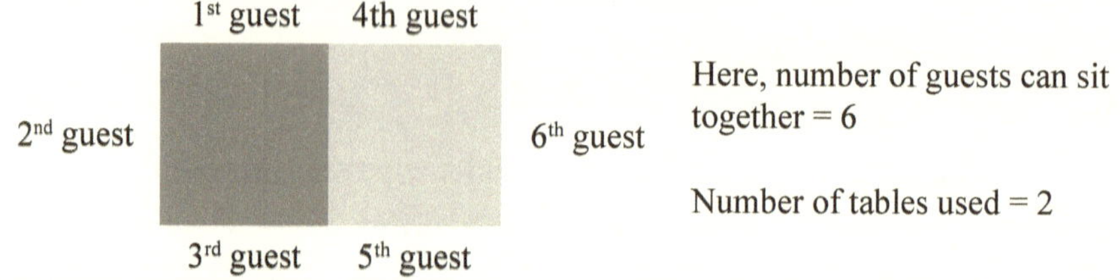

Here, number of guests can sit together = 6

Number of tables used = 2

Arrangement 3

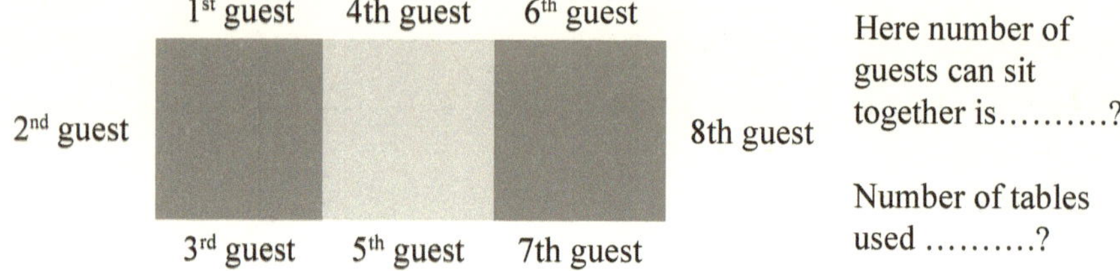

Here number of guests can sit together is..........?

Number of tables used ....?

But Ravi does not know that how many guests will come to his house.

So, can you find that how many tables will be required to sit all guests together?

Now let the number of guests arriving at Ravi's house be **n**

Please fill in the boxes

No. of Table	1	2	3	4	5	6	7	....	n
Guests/ persons	4	6	8	10					?

Now, let us see this

If number of tables is 1 then it is clear that

$1 \times 2 + 2 = 4$ persons can sit together

Similarly,

If no. of tables is 2 then we have

<table><tr><td><u>*Space for Rough Work*</u></td></tr></table>

$2 \times 2 + 2 = 4 + 2 = 6$ persons can sit together

Again, If no. of tables is 3 then we have

$3 \times 2 + 2 = 6 + 2 = 8$ persons can sit together

So, the formula that comes is

No. of tables $\times 2 + 2 = $ **Y** persons can sit together

Therefore,

Let the number of tables used be n

Then we have

n× 2 + 2 = 2n+2 = no. of persons can sit together if n tables were used.

So, we have seen in the above examples that with the use of variables and constant we were able to guess number of tables and number of persons can sit together.

Some examples of equations are:-

a)   x+6 = 2   (variable x)

b)   p+9 =1   (variable p)

c)   2n+2 =10   (variable n)

As, you have learnt about the constant and variable now answer the following questions

Fill in the blanks

a)   In -4x, _____ is constant and _____ is variable.

b)   In 7pq, _____ is constant and _____ are variables.

Polynomials

Polynomial is an algebraic expression in which the variables have non-negative integral exponent. Polynomials can be in one variable or in two variables.

> **_Key point_**
>
> *With the help of an **constant** and **variable** you have created a new thing known as **equation***
>
> *i.e. (2n+2)=m*
>
> *n= no. of tables*
>
> *m = no. of persons can sit together*

> *What is equation?*
>
> *An equation is a statement of an equality containing one or more variables.*
>
> *An equation contains two sides left hand side commonly known as L.H.S. and right hand side commonly known as R.H.S.*
>
> *The study of algebraic equations is probably as old as mathematics: the Babylonian mathematicians*

For example

- $7 + 2x + 3x^2 + x^3$ is a polynomial in one variable x.
- $2x^2 + 3xy + 7$ is a polynomial in two variable x and y.

> **Note:-** In a polynomial the *exponential* value cannot be *negative*
>
> $\longrightarrow$ *Negative value of exponent*
>
> $7 + 2x + 3x^{-2}$ is not polynomial why?

Types of polynomials

Polynomials can be divided into three types on the basis of term it contents

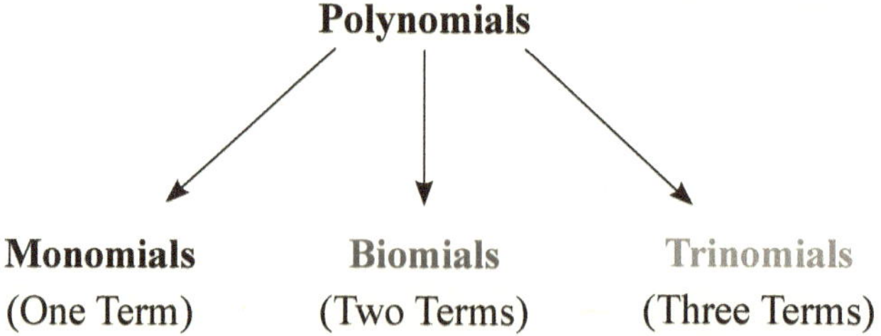

- Monomial is a polynomial of one term for. e.g. 2x, and 7xy
- Binomial is a polynomial of two terms for. e. g. x+2, and 7xy+ 2x.
- Trinomial is a polynomial of three terms for. e. g. $2x^2 + 3xy + 7$.
- And algebraic expression containing more than three terms are called **polynomials.**

💡 Let's Think

Write three algebraic expressions containing more than three terms

1. ....
2. ....
3. ....

> **Note: -** polynomials can have no variables at all e.g. 5 is polynomial having just one term which is constant.

As you all are now familiar with the polynomials, its term and variables. Now let us discuss about the degree of polynomials. Let us consider an example of polynomials in one variable. In a polynomial the highest power of the variable is called the degree of the polynomial

For example:

7x+2 is a polynomial in x of degree one

Let us see here the polynomial 7x+2 has two terms

I term= **7x** and II term = **2**

Now the highest power of variables here is 1. i.e. power of variable x is one

Therefore degree of polynomial is **1.**

Example: $17y^2 + 3y^3 + 6$

Here the term in polynomial is three

I term = $17y^2$, II term=$3y^3$, III term = 6

Variable in the polynomials is y

So, the highest power of variable in polynomials is 3.

Therefore, the degree of polynomial is 3.

Degree of polynomials in two or more variables

If the polynomials have two more variables then, the sum of the powers of the variables in each term is calculated and the highest sum so obtained is called the degree of the polynomials.

Let us find the degree of polynomials in two variables.

$2x^3y - 6y^4 + 8x^4y^2 + 5x^2 + 3$

Here total number of terms in the polynomials is 5.

x, y are two variables present in the polynomials.

Now, in the first term if we calculate the power of both the variables

$2x^3y^1 = 3+1=4$

Powers of variables in II term

$6y^4 = 4$

Power of variables in III term

$8x^4y^2 = 4+2 = 6$ **(Highest sum)**

Power of variables in IV term

$5x^2 = 2$

Therefore, the highest sum of power of variables in all the terms present in the polynomials is 6

So, the degree of polynomials = 6

ASSESSMENT

- **Write in 3xy/k, _____ is constant and _____ are variables**
- **Define polynomials. Write polynomial in one variable.**

....
....
....
....
....
....

- **Identify which of the following expressions are not polynomials**

    a) $x+2$

    b) $4x^3 - 6y^{-5} + 5$

    c) $x^2 + \dfrac{1}{x^2}$

- **calculate the degree of each of the following polynomials**

    1. $(5r^2 + 11r + 6)$

    2. $4x^3y - 6y^5 + 8x^3y^2 + 5x + 3$

Algebraic Expressions

Introduction

In the previous unit, you studied about the basic concepts of algebra like variables, constants and algebraic expressions. The backbone of algebra is an expression formed with the help of variables, around which the whole algebra revolves, and such kinds of expressions are known as algebraic expressions.

> *Algebraic*
> *Expression*
> $2x + 19x + 11 + x^2$

Algebraic expressions are commonly known as the expression. Suppose you have variable x, a constant 7 and a operator '+', with the help of all this, you have to make an expression so, it will be as (x + 7)

Here (x + 7) is an algebraic expression or commonly known as expression.

 Learning Objectives

After going through this unit you will be able to:-

- Define expression.
- Identify the term and co-efficient.
- Perform the addition and subtraction of an algebraic expression.
- Do the multiplication of a polynomial.

Instructional Scheme

To understand an expression it is very necessary to understand some basic concepts of algebraic expression like terms and co-efficient. So let us discuss each concept one by one.

Let's consider the expressions

$$7x + 20 \dots\dots\dots\dots\dots\dots\dots(1)$$

Now, as you know that above expression (1) contains 2 terms. i.e. 7x, and 20 and also contains an operator '+' (**Addition**)

So, here 7x, and 20 are the two terms. That mutually constitutes the expression (1) with the help of an addition operator.

💡 Let's Think

Fill in the blanks

Q.1 $19x + 20 + x^2$

How many terms do the above expressions contains? ___________________

> *A constant term written before or after the variable is known as co-efficient of that variable*

Q.2 $2x + 19x + 11 + x^2$

How many terms do the above expressions contains…………………………..

Co-efficient

Again consider the expression $7x + 20$

Here the first term contain a variable i.e. 7x and here 7 is a co-efficient of x.

Similarly, (-20x) and **(-x)**

In (-20x), **(-20)** is the co- efficient of x.

And in (–x), **(-1)** is the co-efficient of x.

Addition and Subtraction of Algebraic Expressions

(a) Addition of Algebraic Expressions

In the previous classes of arithmetic you have learned about how to do addition like

$$7 + 2 = 9$$
$$2 + 4 = 6$$

These are the simple arithmetic additions in which you add one number to another number by using the arithmetical skills.

But, in the algebra have you ever think that how the addition of unknown quantities

Takes place. For example

$x + x = 2x$ (here x is a unknown quantity)

$2x + x = 3x$

In arithmetic you add as $7 + 2 = 9$

i.e. **1+1+1+1+1+1+1**+1+1 = 9

Here you can say 7 one's and 2 one's added together to get 9 one's.

But, what happens in algebra have you ever think?

Suppose you have to add

$2x + x$

Then it is quite clear that you have to add **two x's** and **one** x's together to get the result

$2x + x$

$x + x + x = 3x$

But when it is $x+y =?$ Then how you will add?

You know that two different variables cannot be added together to get a definite sum i.e.

$x + y$ remains $= x + y$ after addition

So, when you add the same variable with the same variable you actually add the co-efficient of that variable $1x + 1x = 2x$

Again consider two expressions

$7x + 20$ and $3x + 11$

Add both of them

$\Rightarrow$ $(7x + 20) + (3x + 11)$

$\Rightarrow$ $7x + 20 + 3x + 11$

$\Rightarrow$ $(7x + 3x) + (20 + 11)$

$\Rightarrow$ $10x + 31$ is the required addition

Take another example:-

Consider the two expressions

$-5x^2 + 3x + 10$, and $2x^2 + x + 7$

If you want to add both of this expressions then.

$5x^2 + 3x + 10$

$+ 2x^2 + 2 + 7$

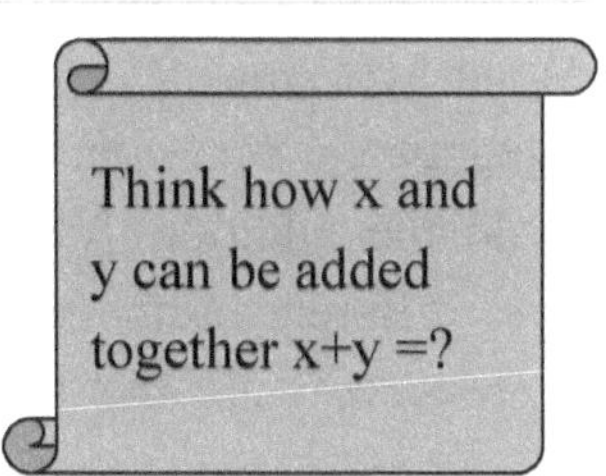

Two different variables cannot be added together

Here you can see
1 + 1 = 2
Coefficient of x's are added together

<u>*Space for Rough Work*</u>

Again here you have three terms in each algebraic expression, since you have to do the addition. You have to add the three different terms of the *first algebraic expression* with the three terms of the *second algebraic expression.*

While adding you have to remember that only like terms can be added together i.e. $5x^2$ can be added to $2x^2$, $3x$ can be added to x, and 10 can be added to 7 only and considering their sign as well.

So, $5x^2 + 2x^2 = 7x^2$, $3x + x = 4x$, $10 + 7 = 17$

(b) Subtraction of Algebraic Expressions

Similarly now suppose you have to subtract these two expression then you have

$$5x^2 + 3x + 10$$

$$(-)\ 2x^2 + x + 7$$

Now for the subtraction the sign of the terms of second expression will be changed due to the minus sign as

$$5x^2 + 3x + 10$$

$$(-)2x^2 + x + 7$$

$$(-)\ (-)\ (-)$$

$$3x^2 + 2x\ 3$$

Here the sign conversion are useful to know which sign has to be performed while addition and subtraction elaborating the above subtraction.

As here again the like terms will be subtracted from the like terms.

- 5x2 and 2x2 are like terms.
- 3x and x are the like terms.
- 10and 7 are like terms.

Here you can see that sign of each term of second expression is changed. i.e.

$2x^2$ has changed to $(-2x^2)$

x has changed to $(-x)$

7 has changed to (-7)

$5x^2$ and $3x^2$

Are like terms

Multiplication of Common Expressions

⇨ $x \times y = ?$

⇨ $x \times x \times x \times x = ?$

⇨ $7 \times x = ?$

⇨ $x^2 \times x = ?$

$$5x^2 - (2x^2) = 5x^2 - 2x^2$$

$$= \mathbf{3x^2} \text{ (sign of } 2x^2 \text{ has been changed to negative)}$$

⇨ Again you have $3x - (x) = 3x - x = \mathbf{2x}$ (sign of x has been changed to negative)

Again 10 - (+7) = **+3** (sign of 7 has been changed to negative)

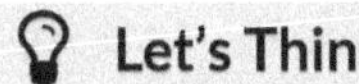 **Let's Think**

Add the following expressions.

(a) $19x + 6$, $5x + 9$

(b) $19x^2 + 14x + 10$, $-9x^2 + 4x - 6$

Multiplication of Algebraic Expression

You have learned to do the addition and subtraction of two algebraic expressions. In this section, you will learn about the multiplication of algebraic expressions. As you know $7 \times 3 = 21$, $9 \times 6 = 54$, but now we will learn how to multiply 2 variables together.

Suppose

$$x \times x = x^2$$

$$x \times x \times x \times x = x^4$$

Here in multiplication the exponent of x's will come into use

Example: - multiply $(2x + 3)(x + 2)$

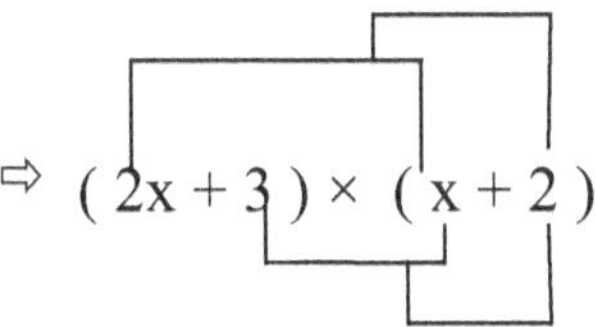

⇨ $2x \times x + 2x \times 2 + 3 \times x + 3 \times 2$

⇨ $2x^2 + 4x + 3x + 6$

⇨ $2x2 + 7x + 6$ is the required solution.

<table>
<tr><td>

2x + 3 has two terms

I term = 2x = I A

II term = 3 = II A

x + 2 = has two terms

I term = x = I B

II term = 2 = II B

When multiplication takes place then

(I A × I B + I A × II B + II A × I B + II A × II B)

</td></tr>
</table>

ASSESSMENT

Q.1 Consider the expression and identify the co- efficient and terms.

$$x^2 + 3x + 2$$

Number of Terms = _____________________

Co- efficient of x 2 = _____________________

Co- efficient of x = _____________________

Q.2 Add and subtract the following expression.

(2x – 3), (x + 2)

Q.3 Multiply the binomials

$(2x - 3) \times (x + 2)$

Q.4 Multiply the binomials

$(2x\ 2 + 7)\ (3x + 1)$

Algebraic Identity

Introduction

In the previous unit, you have studied about the algebraic expression. As well as about the L.H.S. and R.H.S. of the equation, since the expression is said to be true only if L.H.S. and R.H.S. both get equals, but expression becomes true only for some particular value. It does not hold true for all values.

One major question comes into mind, does there exist any expression, that can be true for all values and if exists then what kind of features it will have.

Therefore in the present unit you will learn about such kind of expression. Which are true (L.H.S. = R.H.S.) for all values of variables present in it and they are commonly known as *algebraic identities.*

Algebraic identities are useful in solving many types of questions like solving equations, doing Factorizations etc. algebraic identities are very effective as well as time saving too. Let's see how?

 ## Learning Objectives

After going through this unit you will be able to:-

- Recall the algebraic identities
- Identify and solve the problems by the use of identities

Introduction to Algebraic Identities

Consider an expression

$$7 + x = 3x + 5 \quad \ldots\ldots\ldots\ldots\ldots\ldots\ldots (1)$$

Now, put $x = 1$ in equation 1 you will get

$$7 + 1 = 3 \times 1 + 5$$
$$8 = 3 + 5 = 8$$

Here in the equation
$7 + x = 3x + 5$
$7 + x = L.H.S$
$3x + 5 = R.H.S$

But, if you put x = 2 in equation 1 you will get

$$7 + 2 = 3 \times 2 + 5$$

$$9 = 6 + 5 = 11$$

$$9 = 11$$

L. H. S. $\neq$ R. H. S. Since $9 \neq 11$

So, clearly you can see that equation 1 is satisfied only for only one particular value i.e. x = 1

But, suppose if there may be such type of equation which is satisfied by every values, then which kind of expression it will be? Can you write any expression which is satisfied by all values? ___________________

> **Multiplication of Two Algebraic Expressions**
>
> $(a + b) \times (a + b)$
>
> $\Rightarrow a \times a + a \times b + b \times a + b \times b$
>
> $\Rightarrow a2 + ab + ab + b2$
>
> $\Rightarrow a2 + 2ab + b2$

Let us see an expression

$$(x + 1)^2 = x^2 + 1 + 2x \ldots\ldots\ldots\ldots\ldots(2)$$

Putting the value of x = 0 in (2) you will have

$$(0 + 1)^2 = 0^2 + 1 + 2 \times 0$$

$$1 = 1 \Rightarrow \text{L.H.S.} = \text{R.H.S.}$$

Expression is satisfied

Now put the value of x = 1 in the (2) and calculate?

Does the expression satisfies for x = 1?

Put x = 1 in equation (1) and solve the L.H.S. and R.H.S.

Now putting x = 2 in equation (2)

$$(2 + 1)^2 = 2^2 + 1 + 2 \times 2$$

$$(3)^2 = 4 + 1 + 4$$

$$(3)^2 = 5 + 4$$

$$9 = 9$$

Again the expression satisfies

As **L.H.S. = R.H.S.** $(9 = 9)$

So, it is evident that the expression is satisfied for all the values of x [you can check by putting another value of x in equation (2) like x = 2, 3, 4, 5 etc.

So, finally you have got an expression which is satisfied for all values and such expression those containing a variable and

Which is true for all values of that variable is called an identity.

What is the difference between equation and identity?

As explained in previous section that an equation is true or satisfied when L.H.S. = R.H.S.

For some particular values of variables present in it, but an identity is true /satisfied (L.H.S. = R.H.S.) for all values of variables present in it.

Some Standard Identities

(a) $(a + b)^2 = a^2 + 2ab + b^2$

(b) $(a - b)^2 = a^2 - 2ab + b^2$

(c) $(a + b)(a - b) = a^2 - b^2$

Let's understand it more clearly.

Binomial is an expression which has two terms

What is Algebraic Expression?

In <u>mathematics</u>, *an algebraic expression is built up from integer, constant and variables.*

$$3x + 9 = 12$$

is an algebraic expression

Consider an expression

$$(x + 1)^2 = x^2 + 1 + 2x$$

How?

Let us see (x+1) (x+1)

$\Rightarrow$ **x × x + x × 1 + 1 × x + 1 × 1**

$\Rightarrow$ **x² + x + x + 1**

$\Rightarrow$ **x² + 2x + 1**

Identity I –

$(a + b)^2 = a^2 + 2ab + b^2$

This is an identity which is actually a square of a binomial.

Consider the L.H.S. of the identity.

$\Rightarrow$ $(a+b)^2$

$\Rightarrow$ $(a+b)^2$ can be written as $(a+b)(a+b)$

$\Rightarrow$ $a^2 + 2ab + b^2$ = R.H.S.

 Activity 3.1

Let's verify it step by step

$$(a+b) = a^2 + 2ab + b^2$$

Now putting a = x and b = 6 in the L.H.S. of the equation you have

$$(x + 6) = a^2 + 2ab + b^2$$
$$= x^2 + 2 \times x \times 6 + 6^2$$
$$= x^2 + 2 x \times 6 + 6^2$$
$$= x^2 + 12x + 36$$

> *Multiplying a binomial with binomial*
> *(x+6) (x + 6)*

So yeou have

Now, consider the L.H.S. $\boxed{(x + 6)^2 = x^2 + 12x + 36}$ = $(x + 6)^2$

$\Rightarrow$ $(x + 6)^2 = (x+6) (x + 6)$

$\Rightarrow$ $x^2 + 6x + 6x + 6 \times 6$

$\Rightarrow$ $x^2 + 12x + 36 = $ **R.H.S.**

Hence verified

> **NOTE: In the upcoming unit you will learn the Factorization of trinomials $ax^2 + bx + c$,**
> **and by the use of Factorization you will obtain two factors of $x^2 + 12 + 36$**
> Factor I (x+6)
> Factor II (x + 6)

Let us understand the whole process of product of two binomials in detail with the help of step by step process.

⚙ Activity 3.2

Using the identity solve and verify $(x + 3)$

Let's follow the all step of solving an expression and follow the instruction and fill in the blanks as per requirement.

$$(x + 3)^2 = x^2 + \underline{\quad} + \underline{\quad} \text{ [using the identity]}$$
$$(x + 3)^2 = (x + 3)(x + 3)$$
$$= x^2 + \underline{\quad} + 3x + \underline{\quad}$$
$$= x^2 + 6x + 9$$

Identity II

$$(a-b)^2 = a^2 - 2ab + b^2$$

Verify the identity II using the step explained for verifying the identity I.

....

....

....

....

....

....

....

💡 Let's Think

Solve by using the identities

Q1. $(2x + 3)^2$

Q2. $(x - 11)^2$

Q3. $(m - 7)^2$

Q4. $(2p + q)^2$

Identity III

$$(a + b)(a - b) = a^2 - b^2$$

Here, again consider the L.H.S. of the identity

⇨ L.H.S. = $(a+b)(a-b)$

⇨ $a^2 - ab + ba - b^2$

⇨ $a^2 - ab + ab - b^2$ [since ab = ba then −ab + ab = 0]

⇨ $a^2 - b^2$ = R.H.S.

Hence verified.

> **NOTE:-** (a+b) and (a- b) are two factors of the $(a+b)^2$

 Let's Think

Solve by using the identities $(a+b)(a-b) = a^2 - b^2$

Q1. $(x + 2)(x - 2)$

Q2. $(2x + 1)(2x - 1)$

Q3. $(x^2 - 4^2)$

Use of Identity

Now, in this section you have learnt how to use the identity to solve the various problems like finding the product of binomial expressions. So use the identity as an alternative method to solve the question.

> **Note:-**
> - *With the use of identity increase in accuracy and time saving can be done while solving a question.*
> - *With the use of identity you can factorise the expression as well.*
>   *For example: (a+b) and (a- b) are two factors of the $(a+b)^2$*

⚙ Activity 3.3

Suppose, if you want to find the square of 2 then what will you do?

$=2^2= 2\times2 = 4$

As, here the number 2 is small it is easy to find the square of it. Have you thought that if you have to find the square of 70, 100, 1000, 10999, then what will you do? To find the square of 10999 i.e. $(10999)^2$ you will similarly multiply $= 109999 \times 10999$, but is will be quite time taking process for you. So, let's use the identity to solve such kind of problems.

⇨ Consider an example: $(107)^2$

As you know that

⇨ $107 = 100 + 7$

⇨ $(107)^2 = (100 +7)^2$ [**squaring both the sides**]

Now, consider the R.H.S. $= (100+7)^2$. Does it look similar to the any identities you have studied? $(100 +7)^2$ look likes identity $(a+b)^2$

$(100 +7)^2$ $\qquad\qquad\qquad$ $(a+b)^2$

Now, compare and write

$a =$ _______________

$b =$ _______________

$$\text{Using the identity } (a+b)^2 = a^2 + 2ab + b^2$$

$$(100 +7)^2 = (100)^2 + 2 \times \underline{\quad} \times \underline{\quad} + 7^2$$

$$=10000 + 1400 + 49$$

$$= 21449$$

⚙ Activity 3.4

Find the value of $a^2 + b^2$? When $a + b = 4$, $ab = 10$

Follow the instruction and fill in the blanks.

Here you have to find $a^2 + b^2 =$?

And you have

$$a + b = 4$$

$$(a+b)\,2 = 42 \text{ (squaring both the sides)}$$

Now, as you can see the L.H.S.is a identity so you can use $a^2 + 2ab + b^2 = 36$

$$a^2 + b^2 + 2 \times \underline{} = 36 \text{ [as } ab = 10 \text{ given]}$$

$$a^2 + b^2 + 20 = 36$$

$$a^2 + b^2 = 36 - \underline{}$$

$$a^2 + b^2 = 16 \text{ is the required solution.}$$

🎲 ASSESSMENT

Using the identities solve the following

Q1. $(2a + b)^2$

Q2. $(p - 2q)^2$

Q3. $(m^2 - n^2)$

Q4. $(102)^2$

Exponents

Introduction

In the previous unit you have learnt about the variables, algebraic expressions, identities, and their addition and subtraction. Let's considered an algebraic expression or an polynomial

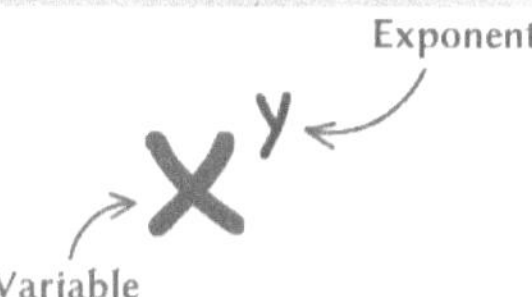

$$x^2 + 2x + 1 \ldots\ldots\ldots\ldots\ldots\ldots\ldots(1)$$

The polynomial contains 3 terms and having degree 2.

Let's us considered the first term of polynomial (1) i.e. x^2.

Here x is a variable and 2 is a constant as well as power raised on x but we can call 'x' as a base and 2 as exponent.

So, in the present unit we will learn about the exponents.

Learning Objectives

In this unit, you will make /get an overview of basic concepts of exponents and introduction to the laws of exponents.

After working through this unit, you will be able to:-

- Identify exponents.
- Define exponents.
- Interpret and identify the negative exponents.
- Recognize the laws of exponents.
- Use the laws of exponents in solving questions.

Instructional Scheme

Suppose you have to multiply 'x' five times. i.e.

$$x \times x \times x \times x \times x$$

But, as the values of x is unknown you cannot multiply it to get a certain values, so you will write it as…………

$$x \times x \times x \times x \times x = x^5$$

So, here x = base and 5 = exponent

Now, if we write product of a number by itself many times such as.

$$3 \times 3 \times 3 \times 3 = 3^4$$

Exponent

Base

The expression 3^4 is read as three to the fourth power also the expression 3^4 or 81 is called a power of 3. Here 3^4 is known as **exponential** from of the power.

Let's us considered a negative number (-1) to the fourth power i.e.

$(-1)^4 = 1$

$-1^4 = -1$

Here have you notice the importance of parentheses around the (-1). Let's do an activity to be much clear about this.

$(-1)^4 = -1 \times -1 \times -1 \times -1$

$= 1 \times 1 = 1$

$-1^4 = -(1 \times 1 \times 1 \times 1)$

⚙ Activity 4.1

Firstly we have $(-1)^4 = -1 \times -1 \times -1 \times -1$

Multiplication of negative term with negative term yields positive term

Multiply first two terms $= -1 \times -1 = 1 \ldots\ldots\ldots$ (a)

Now multiplying the next two terms $= -1 \times -1 = 1 \ldots\ldots\ldots$ (b)

Now using (a) (b) we get

$$1 \times 1 = 1$$

Therefore,

$$-1 \times -1 \times -1 \times -1 = (-1)^4$$

But, if we solve

$$-1^4 = -(1 \times 1 \times 1 \times 1)$$

Then after multiplying it becomes $-(1) = -1$

Clearly, $-1^4 = -(1 \times 1 \times 1 \times 1) = -1$

Hence we get

$$(-1)^4 \neq -1^4$$

💡 Let's Think

Identify the base and exponent in the following.

Q.1 x^y Exponent………… and base……………

Q.2 1 Exponent………… and base……………

Q.3 2^{-7} Exponent………… and base……………

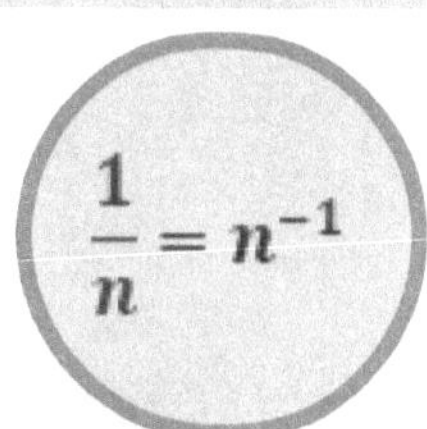

You have learned about the exponent with the positive values. Now, let's discuss about the exponent with the negative values and consider the expression.

$$2^{-2} = 1 / 2^2 = 1/ 4$$
$$3^{-3} = 1/3^3 = 1/ 27$$

Here in the above expressions exponent is negative i.e. -2,-3.

Note:- for any real number $a \neq 0$, and an integer n, $a^{-n} = 1/a^n$

⚙ Activity 4.2

Let's consider the Exponent  Negative

$$a^{-n}$$

Step 1: $1/n = n^{-1}$

$1/n$ can be written as n^{-1}

Step: 2

$$\tfrac{1}{2} = 2^{-1},$$
$$\tfrac{1}{4} = 4^{-1} = 1/2^2 = 2^{-2},$$
$$1/3^2 =?$$

In general we can conclude that $1/a = a^{-1}$

$$1/4^3 = \text{……………………}$$

Or $1/a^n = a^{-n}$

💡 Let's Think

Q.1 $1/? = 2^{-2}$. ? = ...

Q.2 $1/2^0 = $ ....

Q.3 $a^{-n} \times 1/a^n = $...............

Laws of Exponents

There are mainly five laws of exponents. Here you will discuss and learn all of them one by one.

Law I

If $a \neq 0$ be any number and m, n be any integer, then we have

$$a^m \times a^n = a^{m+n}$$

Let us consider an example

$$2^3 \times 2^4 = 2^{3+4} = 2^7$$

Then a $=2$ m $=3$, n$=4$

$$2^3 \times 2^4$$

$$a=2 \ m=3, \ n = 4$$

L.H.S$= 2^3 \times 2^4$ and R.H.S$=2^{3+4}$

Considering L.H.S.

$$= 2^3 \times 2^4$$

$$=2 \times 2 \times 2 \times 2 \times 2 \times 2 \times 2$$

(Total no. of 2's are seven, count it)

Adding the total no. of 2's we get that there are seven 2's present

So, we can write as $2 \times 2 \times 2 \times 2 \times 2 \times 2 \times 2 = 2^7 = 128$

Now, consider the R.H.S.

$$=2^{3+4} = 2^7 = 128$$

L.H.S. = R.H.S

Hence the 1st law of exponent is verified

Find the value of

$2^7 = 2 \times 2 \times 2 \times 2 \times 2 \times 2 \times 2$

....

....

....

....

....

....

....

....

....

Calculate

$(2 \times 2 \times 2)^2$

$= (2 \times 2 \times 2) \times (2 \times 2 \times 2)$

=...

....

....

....

....

Law II

$$(a^{m})^{n} = a^{mn}$$

$$(2^{3})^{2} = 2^{3\times2} = 2^{6}$$

(the exponents are multiplied together

$$L.H.S = (2^{3})^{2}$$

$$= (2\times2\times2)^{2}$$

$$= 2\times2\times2\times2\times2\times2 =$$

(total no. of 2's are six, count it)

=Adding the total No of 2's we get that there are six 2's present

So, it becomes $2^{6,}$ which is equivalent to the R.H.S.

$$2^{6} = 64$$

$$L.H.S = R.H.S$$

Hence, the 2's law of exponent is verified.

Similarly

Hints to verify the law of exponent

I.   Consider an expression in the form $(ab)^{m} = a^{m}b^{m}$

II.   Then solve the L.H.S.

III.   Then solve the R.H.S.

IV.   Compare both the sides

Law III

$$(ab)^{m} = a^{m}b^{m}$$

Verify by using suitable exponent.

....

....

....

....

....

....

Law IV

$$(a/b) = a^{m}/b^{m}$$

Verify by using suitable illustration

....

....

....

....

....

....

Law V

$$a^m / a^n = a^{m-n}$$

Verify by using suitable illustration

....

....

....

....

....

....

ASSESSMENT

Q1. Evaluate

$(300^0 + 50^0 + 1^0)$

Q2. $a^6 \times a^{-3} \times a^{-3} = ?$

Q3. If $2^3 \times 2^{-4} = 2^{(p+2)}$ **Then p=?**

Factorization I

Introduction

In the previous unit, you have learned about the algebraic expression its terms and coefficient. You also know that if two numbers are multiplied together to get a product then those two numbers are called the factors.

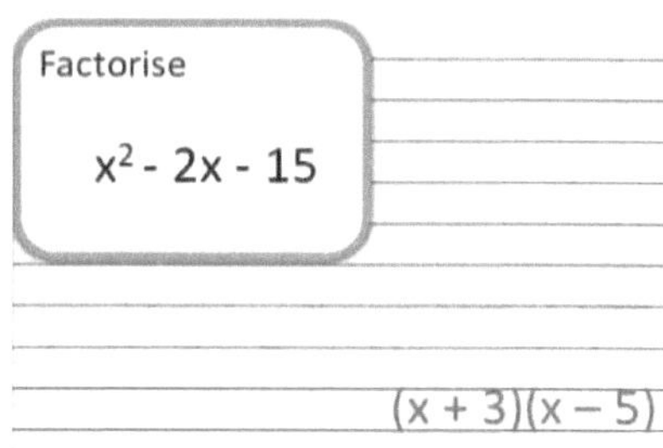

For example: $12 \times 3 = 36$

$= 4 \times 9 = 6 \times 6 = 36$

As you all know in algebra $x(y + z) = xy + xz$

Therefore, you can conclude that Factorization is a representation of an algebraic expression as a product of two or more expressions together.

Learning Objectives

In this unit, you will study about the Factorization, factors of natural numbers, and factors of algebraic expression.

Here an attempt is made to discuss and elaborate the all fundamental concepts of Factorization

After working through this unit, you will be able to:-

- Recognize the monomials factors and factors of the natural number.
- Do the Factorization using common factors.
- Do the Factorization by regrouping terms.

Instructional Scheme

Let us consider the polynomials (xy +xz).

Here, the polynomials (xy +xz) has two factors x and (y+ z).

Similarly, let us consider the natural number:

$$20 = 10 \times 2$$
$$= 5 \times 4$$

Here, 20 has factors 10, 2, 5, 4, 1

and 10, 2, 5, 4, 1 are known as factors of 20.

⚙ Activity 3.1

Consider a brick which we usually use in making houses; you must have seen this brick which is of cuboidal shape. See figure 5.1

Figure 5.1

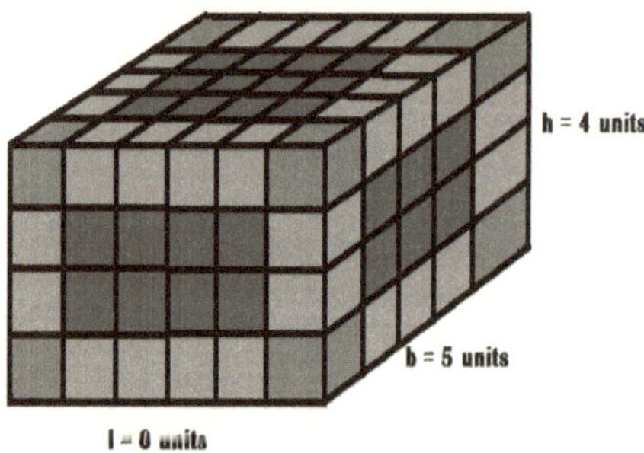

Figure 5.2

Now, write how many faces it have?

....

....

Suppose it has n small cuboids which constitute it as a full cuboidal brick.

Then, you have area of full cuboids = sum of area of all n cuboids.

So can you say that 'n' cuboids figure 5.2 that constitute the full cuboid in figure 5.1 can be treated as a factors of cuboid in figure 5.1

....

If yes then how explain

....

....

....

....

And if no explain how?

....

....

....

....

Test your self:

Find the factors of the following natural numbers.

Q.1 30

....

....

Q.2 40

....

....

Q.3 50

....

....

Factors of Algebraic Expressions

Consider an expression 7pq.

As you know 16pq can be expressed as

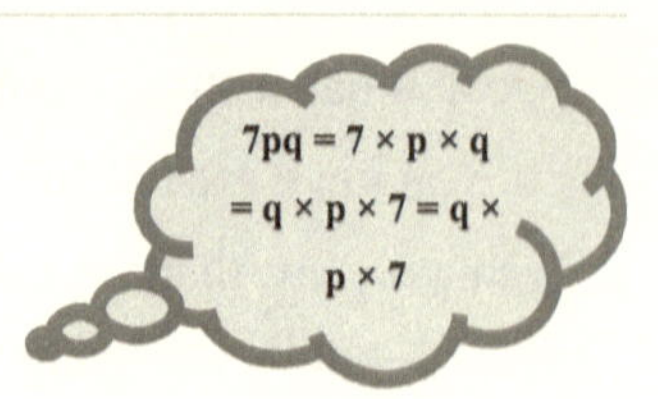

$$7pq = 7 \times p \times q$$

Here it is evident that the 7, p, q are the prime Factorization of 7pq.

Also the expression $7 \times p \times q$ is an irreducible (simplest) form of 7pq.

Next, let us consider the another expression

$$a (b + c + d) = ab + ac + ad$$

Here a and (b + c + d) are the two factors of the expression (ab + ac + ad)

Factorizing a polynomials whose terms have a common monomial factors

Consider an example:

Example: 12x- 12y

Here, if you take out the common factor from both the terms i.e. 12 then the expression will be factories as:

$$12 (x- y)$$

Here again 12 and (x- y) the two factors of the expression (12x-12y).

Factorization by regrouping terms

What is regrouping?

Rearranging the terms in an expression according to the similarity with respect to variables coefficient and degree of the terms to form similar-looking groups is known as regrouping.

How to Regroup?

Let us consider an expression.

$$ax + bx + (-ay) - by$$

Here, the expression contains four terms

$$ax, bx, -ay, -by.$$

Now, rearrange the expression to form a group with similar looking on the basis of variables. If we have Coefficient and degree of terms:

ax- ay + bx –by ....[shifting the terms]

a (x-y) + b (x- y)..[taking the common factors] (x-y) (a+b)

In the above example you have seen that with the help of regrouping, the two factors were obtained i.e. (x-y) (a+b). Hence, the expression is factorized with the help of regrouping.

Test yourself:-

Factories the following expression with the help of regrouping method

Q.1 $x^2 y - x z^2 - xy + z^2$

....

....

....

....

Q.2 $a^2 + bc + ab + ac$

....

....

....

....

ASSESSMENT

Q.1 Factorise

$xyz + xy^2 + yz$

Q.2 Factorise using common factor

70 xyz, 169 pq,

Q.3 Factorise by regrouping the terms

$p^3 + py - q^2 - qyp$

UNIT
06

Factorization II

Introduction

In the previous unit, you have learned about the factorization with the help of common factors and with the help of regrouping terms. There are some other interesting methods of Factorization. So, let us explore some new ways of factorizing the algebraic expressions. Therefore, in the present unit you will learn some new ways of factorization like ***factorization with the help of identities***, and ***factorization of trinomials*** of the form $ax^2 + bx + c$

$x^2 - 2x - x + 2$

$x(x - 2) - 1(x - 2)$

$(x - 1)(x - 2)$

Learning Objectives

After working through this unit, you will be able to:-

- Do the factorization with the help of identities.
- Identify the factors of the form $(x+a)(x+b)$
- Resolve the factorization of trinomials of the form $ax^2 + bx + c$

Instructional Scheme

As you have studied about Algebraic identities and its application in unit 3.

Let us recall all the identities:-

$(a + b)^2 = a^2 + 2ab + b^2$ ·············· Identity I

$(a - b)^2 = a^2 - 2ab + b^2$ ················· Identity II

$(a+b) = a^2 - b^2$ ······················· Identity III

Now, let us consider a polynomial

$$x^2 + 6x + 90$$

Here, if you have to do the factorization of the above polynomial. Firstly you have to identify the identities which you can use in the present factorization, then only you will be able to do factorization with the help that particular identity. Let us understand this step by step.

Step 1

Recognize the identity which suites the best as per the expression

$$x^2 + 6x + 9$$

The expression contains 3 terms, i.e. x^2, 6x and 9

Therefore, it is similar form of $a^2 + 2ab + b^2$

x^2+6x+9

I term = $\mathbf{x^2}$

II term = **6x**

III term =

Step II

Now, after identifying the identity you have to make a comparison with the terms of identity.

$$x^2 + 6x + 9$$
$$a^2 + 2ab + b^2$$

You can clearly see that after comparing the expression with the identity.

$$x = a, b = 3, 2ab = 6x = 2 \times x \times 3$$

Step III

Then finally you will have the expression in the form

$$\mathbf{a^2 + 2ab + b^2 = x^2 + 2(x) \times (3) + 3^2}\ [\text{using the values of step 2}]$$

Since,

$$a^2 + 2ab + b^2 = (a+b)^2$$

So, by comparison $x^2 + 6x + 9 = (x+3)^2$

Therefore, $(x+3)^2$ is the required Factorization.

Activity 6.1

If the area of a rectangle is $x^2 -3x +2$ then find the length and breadth of the rectangle.

Here, the area of the rectangle is given by the expression $x^2 -3x +2$, i.e. area of Rectangle $= x^2 -3x +2$

Now can you write the formula for the area of rectangle?

Area of Rectangle $= L \times$ .... (1)

Factorizing the expression $x^2 -3x + 2$

You have,

$$x^2 -2x - x +2$$

$$x (x - 2) -1 (x - 2)$$

$$(x - 1) (x - 2)$$

Here, you get two factors of the expression

$$x^2 - 3x +2$$

Therefore, you can write $(x-1) (x - 2)$.......... (2)

Now comparing (1) and (2)

You will have

> *Hint*
>
> *Here, you can do the factorization with the help of area of a rectangle. Since,*
>
> *area of rectangle is equal to length × breath*
>
> *and in this case length and breadth are two factors of the expression*

Area of rectangle $= L \times B$

$$= (x - 2) (x - 1) = x^2 - 3x +2 \dots\dots\dots\dots\dots\dots (3)$$

$$L \times B = (x - 2) (x - 1) \dots\dots\dots\dots\dots \text{Using (3)}$$

Clearly you can see that

$$L = (x - 2) \text{ or } (x - 1) \text{ and}$$

$$B = (x - 2) \text{ or } (x - 1)$$

 **Let's Think**

Factorise the Following

Q. 1 $x^2 - 12x + 36$

....

....

....

....

....

....

....

Q.2 $25x^2 - 81$

....

....

....

....

....

....

....

Factors of the form (x+a) (x+b)

In the previous section you have factorise the algebraic expression by the help of identities by identifying their types and using the relevant identities $(a + b)^2$ or $(a - b)^2$ or $(a^2 + b^2)$. There are some other algebraic expression those are not perfect squares, those cannot be compare with identities to do their Factorization. They are similar form of $x^2 + (a + b) x + ab$

Factorising trinomials of the form ax² + bx +c

In this section you will learn, how to factorise a trinomial (algebraic expression which has 3 terms) of the form $ax^2 + bx + c$

Here in $ax^2 + bx + c$.

ax^2, bx, and c are the three terms present in the expression.

Now, you will factorise this expression in the following way:

Let's consider it step by step.

Step I

We have $ax^2 + bx + c$

Coefficient of $x^2 = a$

Coefficient of $x = b$

c = constant term

Here, you have to multiply the product of coefficient of x^2 i.e. 'a' and constant i.e. c so product of coefficient of x^2 and c will be equal to

$$= a \times c = ac$$

$$ax^2 + bx + c$$

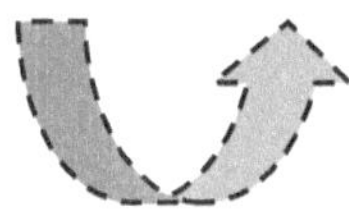

Step II

Now the sum of a and c should be equal to the coefficient of x (second term) in the middle term of the expression $ax^2 + bx + c$

Step III

After getting the two numbers such that their sum is equal to coefficient of x. Proceeding further we get-

$$a \times c = ac \text{ and}$$

$$a + c = b$$

So, we have to find two numbers whose sum is **b** and product is $a \times c$, two such numbers are 'a' and 'c'.

Step IV

We, have

$$ax^2 + bx + c$$

$$ax^2 + ax + cx + c \text{ [Since b = sum of a and c]}$$

$$ax(x+1) + c(x+1) \text{ [Taking common factors]}$$

$$(ax+c)(x+1) \text{ is the required Factorization.}$$

Step V:

Now, solving further if we have

$$ax^2 + bx + c = 0$$

Then using the results of factorization in step IV we have

$$(ax + c)(x+1) = 0$$

$$(ax + c) = 0 \text{ or } (x+1) = 0$$

$$ax = -c \text{ or } x = -1$$

$$x = -c/a \text{ or } x = -1$$

Here $x = -c/a$, $x = 1$ are the 2 zeros of the polynomial $ax^2 + bx + c$.

⚙ Activity 6.2

Factorise- $x^2 + 5x + 6$

Fill in the blanks.

Write the coefficient of x^2.....................and coefficient of x.....................

Then after getting the product you will get 6, and then you have

$1 \times 6 = 6$ equals to 2×3

Now, product of these two numbers 2 and 3 must be equal to the sum of coefficient of.................

Since $2 + 3 = 5$ coefficient of x

Therefore, we have $x^2 + 5x + 6$

$x^2 + 2x + 3x + 6$ [using 5 = sum of 2 and 3]

$x(x+2) + 3(x+2)$

$(x+3)(x+2)$

Is the required Factorization

💡 Let's Think

Factorise $x^2 + 3x - 10$

....

....

....

....

....

🧊 ASSESSMENT

Factorise

Q.1 $2x^2 + 3x + 1$

Q.2 factorise using identities

    a) $x^2 - 16$

    b) $x^2 + 4 + 4x$

Q.3 Factorise the following

    a) $x^2 + x - 2$

Linear Equations

Introduction

This unit focuses on the basic concepts of linear equation. The concept learned in this unit will enhance your ability to solve problems based on linear equations. The linear equation is basically a kind of equation in which sign of equality plays a role of *balance* between L.H.S. and R.H.S. of the equation.

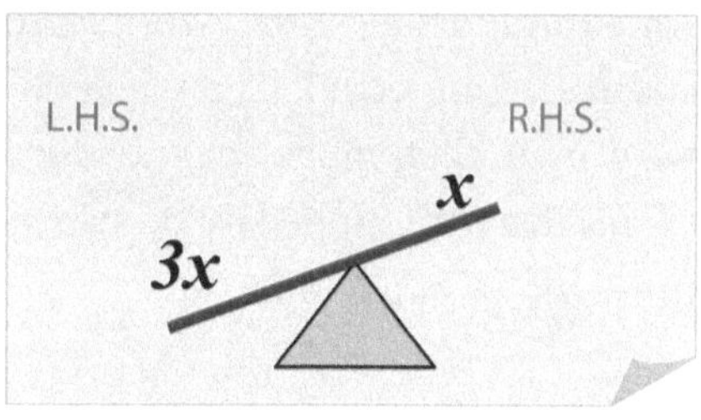

Like $2x + 3 = 5$ there L.H.S. of the equation = R.H.S. of the equation, for any particular values of variable (x) present in it. Also, $ax + b = 0$ is a standard form of linear equation.

 ## Learning Objectives

In this unit, you will get an overview of concepts like equations, linear equations and solving equations. Here, equations will have linear expressions on one side and numbers on the other side of equation. Thus an attempt is made to discuss and elaborate all these concepts.

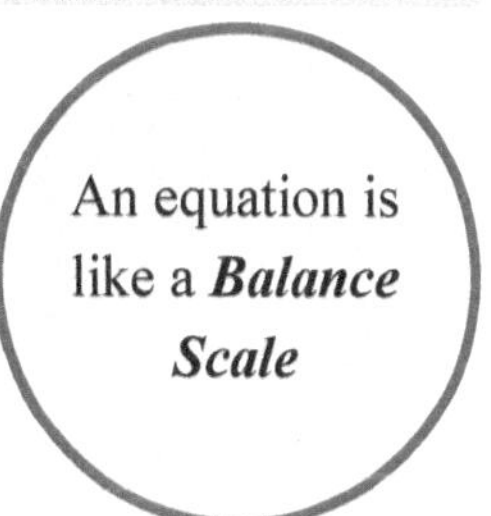

After working through this unit, you will be able to:-

• Recognize the equations.

• Solve the equations.

• Solve the equations which have linear expressions on one side and numbers on the other side.

Instructional Scheme

In general, a **first-degree or linear equation** in one variable is an equation that can be written in the form

$$ax + b = 0$$

Where, "*a*" is not equal to zero. This is called the **standard form** of the linear equation.

An **equation** states that two quantities are equal. This means that the L.H.S and R.H.S are equal. The equation may contain an **unknown quantity** that we want to find. In the equation $5x + 10 = 20$, the unknown quantity is x. This means that 5 multiplied by 'x' which is an unknown quantity that is then added to 10 will be equal to 20.

Steps for solving a linear equation:

Let's consider a equation $ax + b = c$

Step1. Shifting the variable on the one side of the equation:

In the first step your main aim is to get the variable on the one side of the equation.

$$ax + b = c$$

$$L.H.S = ax + b \text{ and } R.H.S = c$$

Here, you have only one variable 'x'.

So it's already on the one side of the equation i.e. L.H.S.

Step2. Transposition

In this step you have to make the transpositions by changing their sign as well.

$$ax = c - b \text{ [transposing 'b' from L.H.S. to R.H.S.]}$$

$$x = (c - b) / a$$

Step3. Substitute the obtained value of 'x' back into the equation to check the solution.

$$ax + b = c$$

Using the value of 'x' you will get

$$a \times (c - b) / a + b$$

$$c - b + b \text{ [negative b and positive b will be cancelled]}$$

$$c = R.H.S. \text{ hence verified}$$

⚙ Activity 7.1

Solve for x: *and fill in the blanks*

Q. $4x - 10 = 42$

$4x - 10 = 42$

$4x = 42 +$..........

$4x =$ ....

$x =$ ....

> <u>*Key Point*</u>
>
> *An equation is a statement in that two mathematical expressions are equal.*
>
> *For example, $2x - 3 = 5$ is an equation containing variable x.*

💡 Let's Think

Q. Solve for *x*

a) $7x - 3 = 5$

....

....

....

....

....

....

....

....

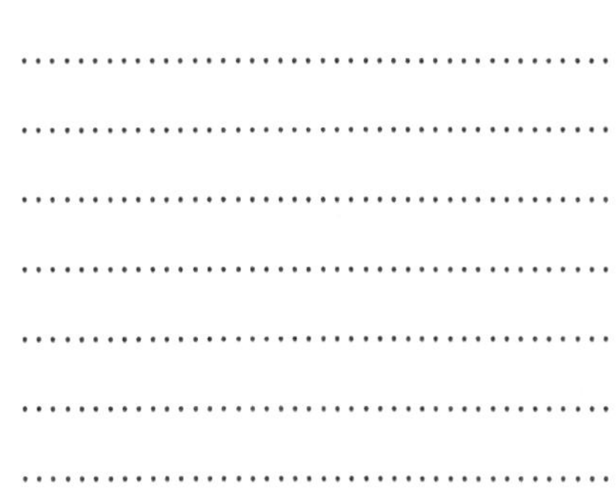

⚙ Activity 7.2

Undoing and Balancing the Equation (Reversing the Process)

Now, as you are familiar with the steps of solving the linear equation. Let's think in a *reverse way.*

For Example: $x - 7 = 9$

There are two ways in which we can proceed in a reverse way

1.   As, it is clear that in the above equation if you add 7 on the both side then you will have. $x - 7 + 7 = 9 + 7$

$x = 16$

2.   Now using the value of $x = 16$ in the equation you will have

 x- 7 = 9

 16 – 7 = 9

 9 = 9 hence verified (L.H.S. = R.H.S.)

 Let's Think

Solve for x

Q. 1) Seven less than x equals 10

....

....

....

....

....

....

Q. 2) Four less than a number multiplied by two

....

....

....

....

....

....

....

ASSESSMENT

Solve for x

Q. 1) $x + 12 = 10$

Q. 2) $3x + 10 = 29$

Q. 3) $17x + 6 = 40$

Applications to Linear Equations

Introduction

In the previous unit, you have studied about the linear equation in one variable.

Also in unit- 2 and unit -3 you have studied about the algebraic expression and it's Factorization. You have also learned about the method of solving equations which have linear expression on one side and numbers on the other side of the equation. Now, in the present unit, you will learn about the application of linear equations.

$$7x - 1 = 2x + 3$$
$$7x - 2x = 3 + 1$$
$$5x = 4$$

You have to use linear equations in your daily life problems like area-related problems, age-related problems, speed and time-related problems.

Learning Objectives

In this unit, you will mainly deal with the linear equations having the variable on the both sides and applications of linear equations related to some specific areas as explained earlier.

How to solve linear equations when there are variables on both the sides of the equations? How to form linear equations from words problems? Therefore, after working through this unit you will be able to:-

- Recognize and solve linear equations having variables on both sides.

- Interpret the word problems based on linear equations.

Instructional scheme

In the previous unit you have studied about the equations in which R.H.S (right hand side) is only a number, but in other situations when R.H.S. can be expression or variables. As in an equation '=' **(equality sign)** plays the role of the balance between the two sides (L.H.S and R.H.S)

As earlier you have seen the equations like:

$$7x-1 = 2$$

Here as you know

$$7x-1 = L.H.S \text{ and}$$

$$2 = R.H.S$$

But, now let us consider the equation

$$7x-1 = 2x + 3$$

Here

$$L.H.S = 7x-1 \text{ and}$$

$$R.H.S = 2x + 3$$

So, it can be seen that this kind of equations which have variables on both the sides.

Algorithm to solve the linear equations having the variables on both the sides_

Key note: To solve a word problem of a linear equation we represent unknown quantities by x and establish the given relation, to construct a linear equation in x. Depending upon the situations whether it will be linear equation having the variable on one side or in both the sides.

Let us consider an equation

$$7x -1 = 2x + 3$$

$$\text{Here L.H.S.} = 7x -1$$

$$R.H.S. = 2x + 3$$

L.H.S contains two terms 7x, and (-1).

R.H.S contains two terms 2x and 3.

Here, firstly you have to do transposition.

$$7x -1 = 2x + 3$$

$$7x - 2x = 3+1$$

$$5x = 4$$

$$x = 4/5 \text{ [Required Solution]}$$

Activity 8.1

Two rectangles are given both having the same perimeter but different measurement of length and breadth given as:-

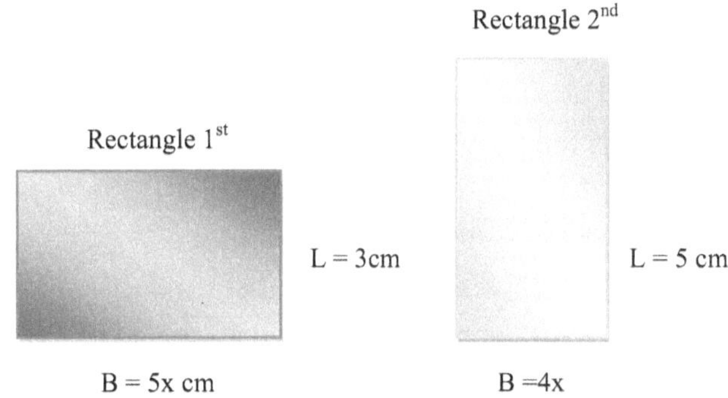

Find the breath of each rectangle?

As you know the perimeter of rectangle = L+ B + L + B

Therefore, the perimeter of 1st rectangle is = 5x +3 + 5x + 3

Similarly, write the perimeter of 2nd rectangle = 4x +............. +............+ 5

Now,

5x +3 + 5x + 3 = ... + 10 [since the perimeter of each rectangle is equal]

So, we have

$$5x +3 + 5x + 3 = 8x + 10$$

$$10x + 6 = 8x + 10$$

$$10x - 8x = 10 - 6 \text{ [transpositions of 8x and 6]}$$

$$2x = 4$$

$$x = ... \text{ [Required solution]}$$

Now, what is the breath given for both rectangles in figures:-

For 'rectangle 1st' B =.................And for "rectangle 2nd" B=......................

So, B = 5 × .... = ... [Rectangle 1]

And B = 4×........... = ... [Rectangle 2]

Application of Linear Equation: Word Problems

The major use of linear equation is its application to solve day to day problems, like problems related to the daily life of area and other related problems of Mensuration, speed and distance related problems.

What is Word Problem?

Word problem is a kind of situation in which the problem is theoretically explained and based on the logic that links one situation to other situations.

It Contains the Linkage of Statement

The major characteristic of word problems is that it relates two or more than two statements together with common linkage between them.

Formation of an Equation

In a word problem, formation of an equation with respect to the given terms and conditions is a major task. The main aim in a word problem is that to construct an equation which can have variable on the both side of the equations.

Solving the Constructed Equation to Get the Solution

This is a final step of solving a word problem. Solving the linear equation formed to get the required result.

Let us consider an example of a word problem and understand it by solving step by step.

> **Example:** The denominator of a rational number is greater than its numerator by 4. If the numerator Is increased by 10 and denominator by 2 the number obtained is 5 by 6. Find the rational number.

Solution:-

Step 1

Understanding the word problem here, it is very evident that you have to find the rational number.

Summarizing the given and demanding situation of word problem as follows:-

(a) Here you have to find a rational number.

(b) In the given rational number denominator is greater than numerator by 4.

(c) In the given rational number if the numerator is increased by 10 and the denominator by 2 then it is equal to 5 by 6.

Step 2

Using (a) from Step 1 let the numerator of rational number be x then by using (b) you have denominator = x + 4

Then required rational number will be as = N/D = x/ x+4……….. (1)

Now using condition(c) from step 1

New rational number as per the condition (c) will be N + 10/ D + 2 = 5/ 6

$$x + 10/(x + 4) + 2 = 5/6$$

$$x + 10/ x + 6 = 5/6$$

Step 3

Now, forming of an equation as per the given condition and solving

$$6(x + 10) = 5(x + 6)$$

$$6x + 60 = 5 x + 30$$

$$6x - 5 x = 30 - 60$$

$$x = -30$$

Then using the value of x = - 30 in (1) we get

x/x + 4 = -30/ -30 + 4 = -30/-26 =30/26

is the required rational number

💡 Let's Think

Q.1 Half of a number is 3 more than one third of the number. Find the number

....

....

....

....

....

....

....

....

....

....

....

....

Q.2 solve for x

2x-3/6 = 0.5x- 3/3

....

....

....

....

....

....

....

....

....

....

....

....

ASSESSMENT

Q.1 Find the number if the number decreased by 2 is 20.

Q.2 The sum of two numbers is 30 and one of them is two greater than the other. Find the numbers.

Q.3. Length of a rectangle is 8 cm more than twice its breath. If the perimeter of the rectangle is 60 cm. Find its length and breadth.

Introduction to Graphs

Introduction

We always require some measuring unit to measure the distance, but distance is measurable if only two fixed points are there, for this location of two points in a space is very necessary. In the other words, we can say that to locate a point or to get a location of a point we need a fixed reference point known as the origin. A French mathematician **Rene Descartes** developed a method to locate a point in the free space.

So, in the present unit you will learn about the graph and locating the points on a graph.

Note: - Distance which cannot be measured is known as infinite distance.

 **Learning Objectives:**

The Present unit deals with the position of a point on a graph and how you will locate various points on a graph. After going through this unit you will be able to:-

(a) Define coordinates and quadrants.

(b) Identify the various graphs like bar graphs, histograms and pie charts.

Instructional Scheme

In the given plane, point O is called origin it is represented by the ordered pair (0,0) here the horizontal line XX' is called x- axis and vertical number scale is called y- axis and the intersection point of both the lines is called origin.

So, you can see that this two number lines horizontal and vertical axes mutually combined together to form co-ordinate Axes.

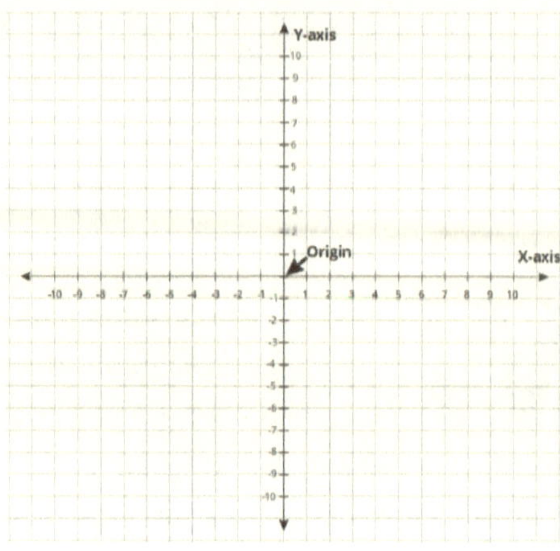

Figure 9.2

> **Note:-** Axes is plural of Axis

Abscissa and Ordinate

Here, in the figure 9.3 the distance OA is equal to the erpendicular distance CB. Here CB is equal to the distance OA.

Now distance from y axis is known as X coordinates or abscissa also, the perpendicular distance of point from the x-axis OX is called the Y coordinates or **ordinate**.

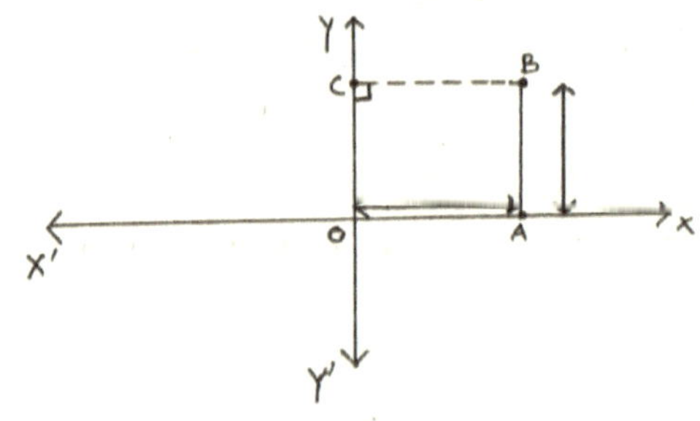

Figure 9.3

> **Note:-** x- axis, y- axis, abscissa, ordinate and origin these all mutually make Cartesian plane.

⚙️ Activity 9.1

Consider the figure 9.4

Suppose Ravi and Surya are two friends living in the different localities but studying in a same school. House number H_1 is Surya's house number and H_2 is Ravi's house number and there school S_1 is located on a Cartesian plane.

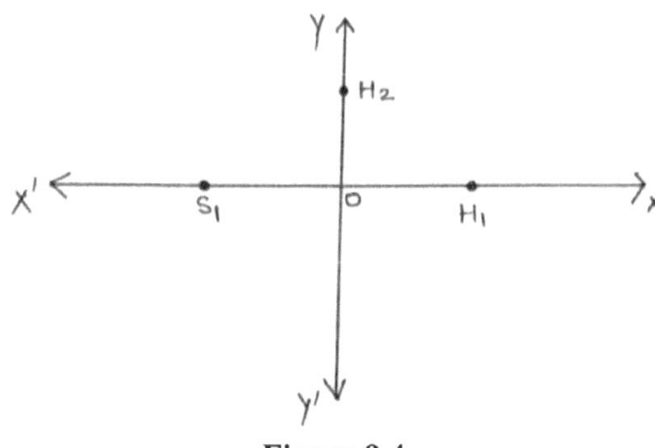

Figure 9.4

Now, can you write on which axes:-

- The school of Ravi and Surya is located……………………………?
- On which axis the house of Ravi is located…………………?
- Which axis house of Surya is located…………………?
- While going to the school at which point Ravi and Surya will meet together………………………………?

💡 Let's Think

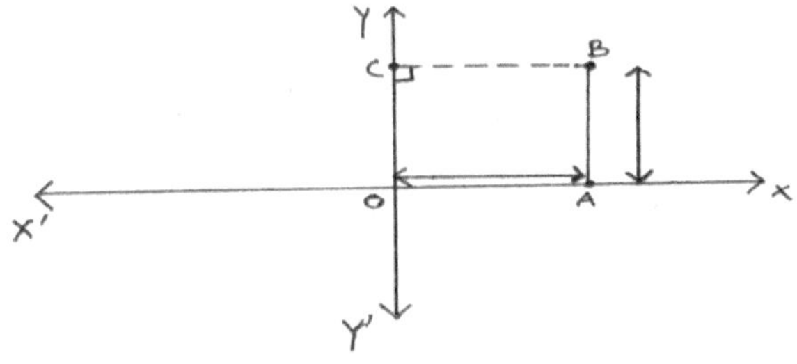

Figure 9.5

Fill in the blanks

Q.1 The perpendicular distance of a point from the y axis is called ……………………….. or………………………………..

Q.2 The perpendicular distance of a point from x axis is called…………….. or ……….

Quadrants

If both the axes, x-axis and y-axis extended on the both sides and full Cartesian plane has been developed. Then as it is very clear from the figure that axes divides a Cartesian plane in 4 equal parts.

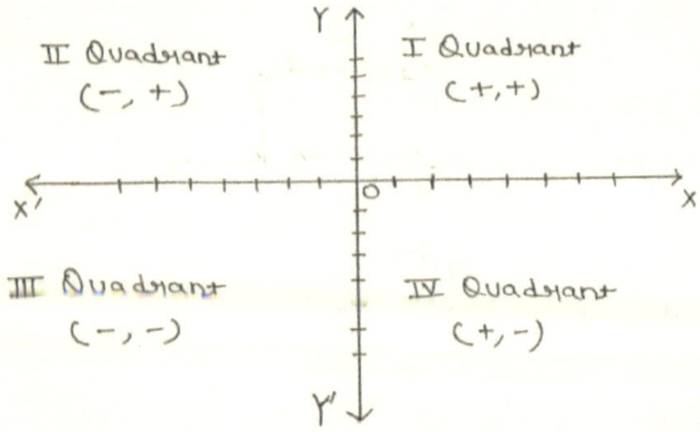

Figure 9.6

See in the figure 9.6

XOY plane = First quadrant

X'OY plane = Second quadrant

X'OY' plane = Third quadrant

XOY' plane = Fourth quadrant

Sign Convention on Four Quadrants

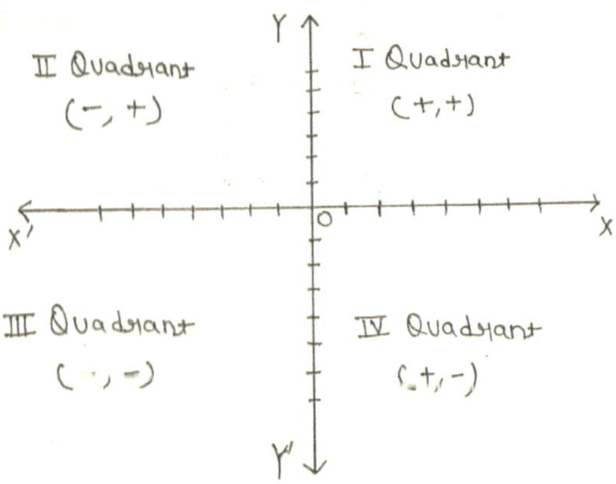

Figure 9.7

Keynote

The ordinate of any point on x-axis is zero, hence the coordinates of any point on the x axis (x, 0) and similarly abscissa *of any point on the y-axis is zero. Thus the coordinates of any point on the y-axis (0,y)*

Here, let us understand it with a ***"trick"***.

(x, y) this is an ordered pair representing

the abscissa and ordinate.

(x, y) = (+, +) means abscissa is positive and ordinate is also positive

(x, y) = (-, +) means abscissa is negative and ordinate is positive

(x, y) = (-, -) means abscissa is …………and ordinate is …………

(x, y) = (+, -) means abscissa is …………and ordinate is …………

⚙ Activity 9.2

Look at the figure and follow the directions to answer the questions given below:-

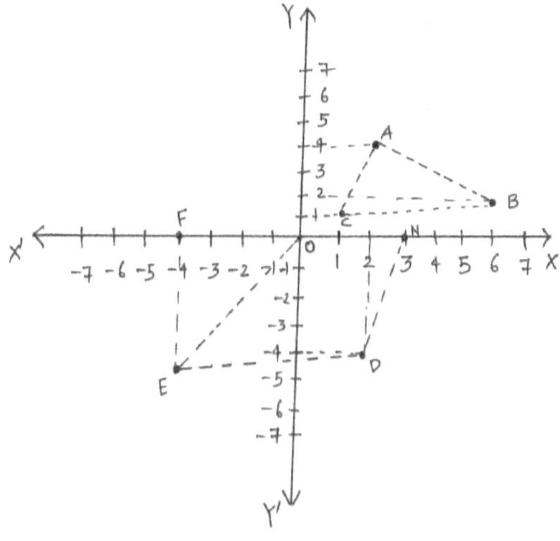

Figure 9.8

How many points do you see on the Cartesian plane given in figure …………………

Write the names of point in the blank spaces

A…….., ……………., D,……………., …………… G,…………,, ……….

Join the points A to B and B to C then C to A.

Name the figure which is formed after joining the points……………………………

Write the coordinates of A, B and C.

Types of Representation on graph

You have already studied about the points on graph and ways to locate it on the graph but there are some special types of graphs plotting which are used to represent the data like bar graph, pie graph, histogram and line graph. Here we will discuss about this type of representation.

A Bar Graph

Suppose that you have to compare the temperature of three different days. Then it is very sure that you have to compare the rise or fall in temperature on three different occasions.

Suppose temperature on the first day is 37 degree Celsius.

Temperature on second day is 28 degree Celsius.

Temperature on third day is 32 degree Celsius.

Then with the help of bar graph you can compare the difference in temperature.

Let us see:-

Note: Depending upon the situation, bar graphs can have double bar as well as single bar also.

Single Bar Graph

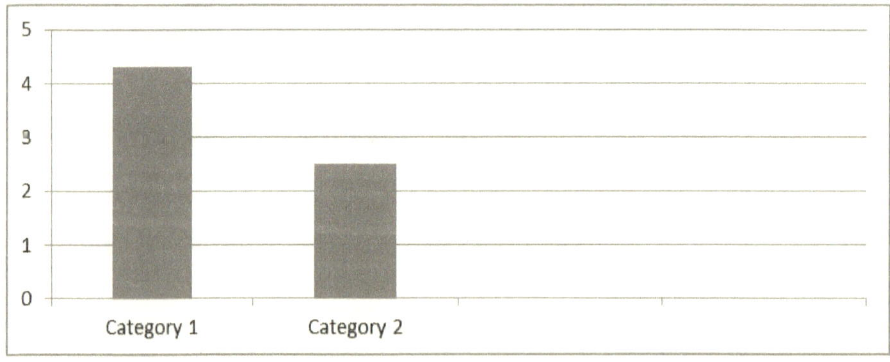

Figure 9.9

Doubled Bar Graph

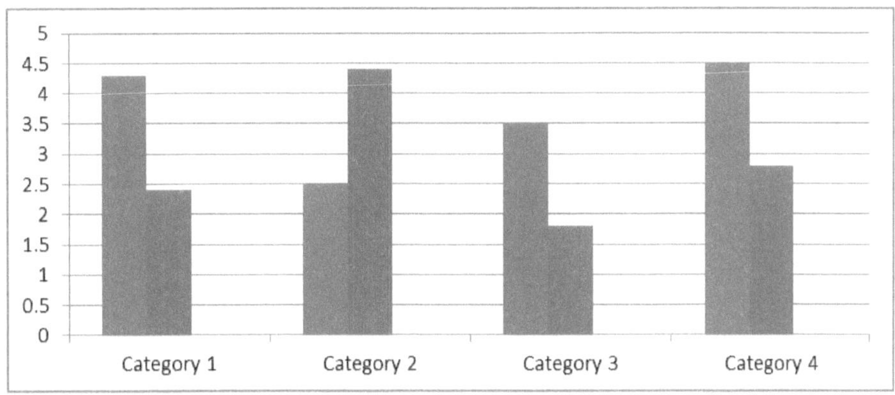

Figure 9.10

Histogram

In the histogram the interval is continuous and similar to the bar graph as shown in the figure below:-

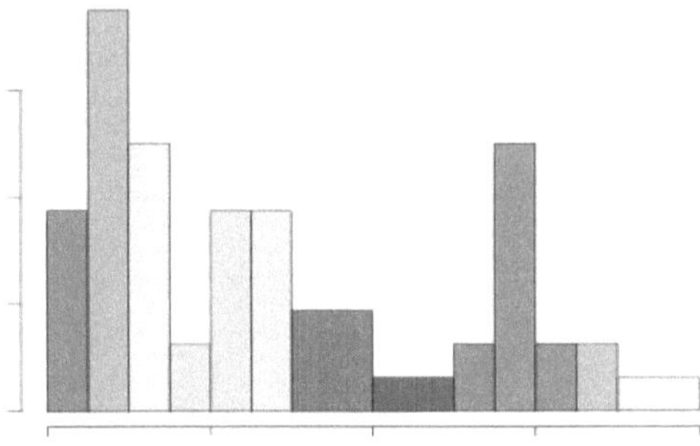

Figure 9.11

Pie - Chart

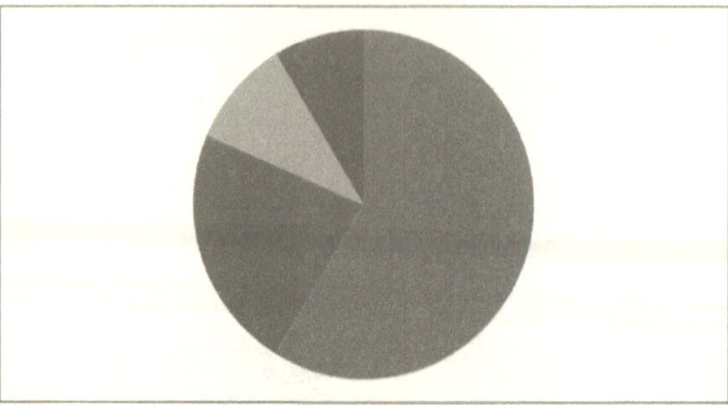

Figure 9.12

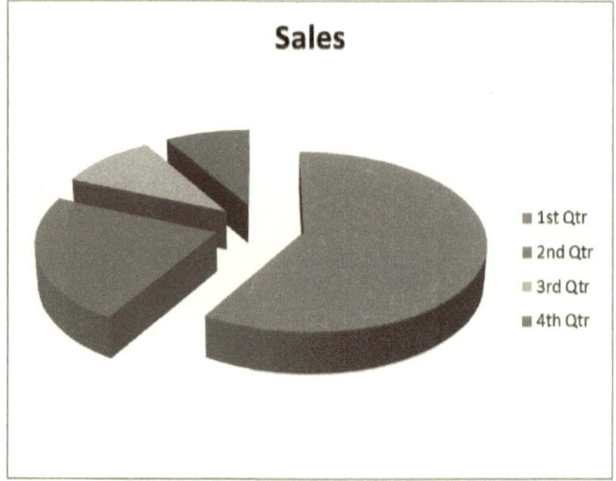

Figure 9.13

Here in the figure 9.12 and figure 9.13 you can see that how different items can be simultaneously represented on a single chart representing their quantities and their shares they have.

Q.1 Let us consider the blood pressure of patients in several ages and draw the appropriate graph based on the given data:

Age group	30-35	35-40	40-45	45-50	50-55	55-60
Number of patients	1	1	4	7	9	12

Q.2 Draw the bar graph using the given data:-

Marks	70	50	60	55
Test	T1	T2	T3	T4

*****

About the Authors

Dr. Prateek Chaurasia is working as an assistant professor, in the Department of Education, Mizoram University. He has also served at the Regional Institute of Education, NCERT, Bhopal, M. P. as an assistant professor. Dr. Chaurasia has completed his Master in Education (M.Ed.) from the Regional Institute of Education, NCERT, Ajmer, Rajasthan and is a recipient of a Gold medal in M.Ed. He has obtained his Ph.D. from Banaras Hindu University, Varanasi, Uttar Pradesh. He is a recipient of prestigious fellowships like the UGC junior research fellowship & NCERT Doctoral Fellowship. He is working in the field of mathematics education, educational technology and assessment in education. He has published many research articles in national and international journals of repute. He is leading the major and minor research projects funded by premium institutions like ICSSR, Mizoram University and the Regional Institute of Education. He has also contributed to various government projects in the capacity of resource person.

***

Dr. Somu Singh has been working as Senior Assistant Professor in the Faculty of Education of Banaras Hindu University for the last 14 years. His expertise is in the field of teacher education, Innovative pedagogical methods & practices, use of ICT in education and Indian knowledge systems. Under his guidance, four researchers completed their Ph.D. Degree and 30 students have completed PG level dissertations. He has published many research articles in national and international journals of repute and many chapters in edited book by national publishers.

He is leading the major research projects funded by premium institutions like ICSSR and Institution of Eminence (IoE) Scheme of the Ministry of Education, GoI. His videos and audio lectures have been recorded and broadcast on Government of India's Swayamprabha programme (Group of TV channels). He has presented many papers and delivered resource lectures in various national and international seminars and conferences. He is the founder of an academic group named 'Teacher Educators for Social Services (TESS)' which has made innovative academic contributions in the field of teacher education during the Covid pandemic crisis, which has more than one lakh teacher educators, students and school teachers as members from all the states of India.

***